DIRECT-POD

THE PRINT

Our guide to self-publishing and book printing

Contents

DIRECT-POD

THE PRINT ON DEMAND COMPANY

Established with 14 years of trading

About us

From small beginnings, Direct-POD has expanded over the years to become one of the most forward thinking book printing companies. We're proud of our people and of the work we produce. At Direct-POD, we build business by providing people with the most effective, sensibly priced printing solutions available. Our customers tell us that's what they want - affordable book printing from a supplier they can trust.

We thrive on even the shortest of deadlines. Our digital presses have a considerable output rate and print at 1200 dpi. We can bind 2000 books a day and our on-line quoting system offers instant quotes, and file upload.

So when you present us with the challenge of a tight turnaround, you'll find we respond quickly and enthusiastically.

Contact information

Direct Print On Demand Ltd
Saxon Fields
Old Harborough Road
Brixworth
Northampton
NN6 9BX

Telephone 01604 882588 Fax 01604 882595

www.direct-pod.com
e-mail: sales@direct-pod.com

Registered in England No. 06860419
VAT Number 970 6130 30

Self Publishing

Introduction

If you've ever thought about publishing your own book, then you have most definitely come to the right place. The art of self publishing is now more practical and affordable than ever before. If you've had a problem trying to get your book published and/or promoted through a mainstream publisher, then self publishing might be the answer you're looking for and the solution to all of your publishing related troubles.

We are here to help you through the process of self publishing, to deal with everything that could give you a problem and allow you a much swifter and easier route to publishing success. We can provide all the information you need to set up your own self publishing scheme and get on the fast track to earning what you deserve, thanks to your own hard work.

Self publishing is something that can be done well by anyone, you just need the determination and the will to go through with it, which the information and examples on this website should help you with. It might sometimes be a bit of a challenge, but you will certainly reap the rewards and benefits that self publishing has to offer.

Simply put, self publishing means that you are the one in control of how your book is published. You get all the copyright, rather than some fancy guy in a suit taking everything that belongs to you and the best thing is that you get to keep one hundred percent of all the royalties that are related to the publishing of your book. You decided when and where it's published, so you really do have complete control over your own hard work and can shape the direction it heads in.

This road is relatively easy, so don't be put off by people telling you that you can't do it or don't have enough money or something like that. Self publishing is something that anyone can do, it's something that's easy to access and control, as it's YOU that has total control over what happens to the thing you want to publish, we act in accordance to your will, providing you with a gateway to better things.

To start with, it might seem like a daunting task, but with our guidance and advice, you'll hopefully be a successful self-published author of an incredible piece of work in no time at all!

Benefits of our services

Unlike many other POD companies we pride ourselves on our flexibilty and the ability to help produce your book just as you want it regardless of format or run length.

- Soft and Hard back options available
- Any size undertaken
- Colour books and colour sections available
- Fast turnaround
- High quality 1200 dpi mono and colour printing
- Various papers available
- Quotes, payments and file upload available online
- Reprint orders can be made with ease
- Secure file storage
- Dedicated customer service team
- Assistance for authors
- Quality hard and soft back books
- Pricing to suit every budget
- Experienced friendly staff
- 1 to 2000 copies
- Colour Books and colour sections available
- Black and White books (with colour laminated covers)
- Free no obligation quotes
- Free advice on how to self publish
- Book Production Specialists
- Staff with over 25 years experience
- BPIF Members
- Investors in people
- Enviromental printers
- Approved Oce quality centre
- UK-USA-Australia POD links

Glossary of book terminology

Basis weight of paper – The weight of a material defined by g/m2 (grammes per square metre).

Bleed – Printed matter which runs off the edge of the paper. The bleed is the amount by which the illustration extends beyond the trim size (We require 3mm of bleed).

Blocking – Impressing a design or lettering on a book cover, the blocking may be in ink or metal foil, or it may be blind blocked to produce a recessed surface.

Book-block – The pages of a book when prepared for casing in, before the case is added. Also book pages trimmed all round ready for punching.

Brass – A brass die used for blocking. Bulk – The thickness of a book or sheet of paper.

Case – The cover of a book, prepared beforehand for affixing to the book-block.

Casebinding – A method of binding in which the cover is made separately but consists of rigid or flexible boards covered with paper, cloth or other material, in such a manner that the covering material surrounds the outside and the edges of the board. Covers always project beyond the trimmed edges of the next pages.

Casing in – Inserting the book block into its case and pasting down.

Collate – To check the sections of a gathered book to ensure that they are all present and are in the correct order.

Cut – The act of cutting flat printed or unprinted paper on a guillotine or three knife trimmer.

Draw on cover – A paperback cover which is attached to the book block by gluing the spine.

Edges – The three cut sides of a book-block.

Endpapers – A piece of plain, fancy printed paper folded into four pages to the page size, sewn or pasted to the first and last section to secure the book to the case or binding.

Extent – The length of a book in terms of the number of pages.

Flyleaf – The inside leaf of a four page endpaper which is pasted along the folded edge of the first or last page of the book.

Foil – Sized metallic or pigment leaf used in blocking lettering or designs on the surface. Used primarily for blocking book covers.

Foredge – The outer edge of a book, opposite the spine.

Foot – The bottom edge of a book.

Frontispiece – An illustration on the page facing the title page of the book.

Head – The top cut edge of a book.

Jacket – The printed paper outer cover which wraps round a case bound book in order to protect it and to act as a point of sale display.

Laminating – The application of a film, either gloss or matt finish, to the surface of the printed book cover to enhance its appearance and to increase durability.

Landscape – The Shape of an illustration or book is referred to as 'Landscape' when its width is greater than its height.

Leaf – A sheet of a book containing two pages, one on each side. Thus, a book containing 32 leaves has 64 pages.

Limp cover – A flexible book cover from 240gsm (Standard) up to 300gsm in weight distinct from a board cover.

Perfect binding - A style of unsewn binding in which the leaves of a book are held together at the binding edge by adhesive.

Portrait – The shape of a book or illustration is referred to as portrait when its height is greater than its width.

Saddle stitching – A method of binding one or more sections with or without a cover by means of wire staples through the centrefold.

Section – A sequence of pages forming part of a book.

Self Cover – A booklet cover of the same paper as the inside leaves, generally the first and last pages on the sheet.

Side gluing – A line of adhesive on the front and back of a book-block to strengthen the hinge opening.

Spine – The bound edge of a book, also known as the 'back'.

Spiral Binding – Attaching single sheets by means of a wire or plastic coil passed through punched holes.

Square back – A binding which has been collated and trimmed but not round and backed. Also called 'Flat Back'. 12

Manuscripts

Manuscripts can be submitted in various formats, but the safest for replication is PDF which is a more secure way of making sure it prints as you see it on the computer screen. If you supply a word file please make sure that it is set up at the correct finished page size, and with the margins you feel happiest with. We can format your manuscript for you but this incurs an extra charge, based on £25 per hour. We also have a service where you can submit an original book to be scanned and reproduced, or hard copy that needs scanning from a typed manuscript with no electronic files. If it comes in as a hard copy that needs OCR scanning, again this is possible but would incur an extra charge for which the price would be subject to sight.

Timing

- We usually aim to get proofs to our clients with in 2-3 working days, and from proof approval delivery should be 5-7 working days.

- If you require the books for a set date please advise on order placement and we will do all we can to achieve that date.

Page layout/format

- We can accept your manuscript in most PC-based software formats but it must exist as one single document. Please do not submit each chapter or each page separately.

- We would print the pages exactly as you supplied them unless you wanted us to carry out the design work for you, for which there would be an additional cost.

- There are no rules about the font you use or the size; those are completely your own choice. We generally use a font called Times New Roman with an 11 point size. A margin of around 15mm should be left around each page, although the choice is yours.

- For best quality results supplying your manuscript in PDF format is ideal. If you do not have the software to convert your document it is possible to get it converted for no cost on various internet sites. (Type "convert to pdf free" into any search engine to find one). If you have any problem with this we can do it for you from many PC based softwares at an additional cost.

- It is generally accepted that odd-numbered pages should be on the right-hand side as you look at the open book.

- A new chapter starts on a new page. (Do use a page break – Alt I B P in Microsoft Word – rather than a series of returns to achieve that).

- Some people prefer new chapters to start on a right-hand page which may entail inserting a blank page beforehand.

- If you want a gap between each paragraph, insert automatic spacing to appear after each paragraph – Alt O P E in Microsoft Word.

- Be wary that a computer spell checker is no guarantee of correct spelling, and the language setting on your PC accounts for many problems here.

- Do not manually hyphenate words at line breaks. Let your word processing application do that for you automatically. (Alt T L H in Microsoft Word)

- The most common mistake is the over-use of capital letters, which should be reserved for proper nouns (names) and the first word in each sentence only. Too many people write sentences such as The Doctor wrote to the Council to complain about the noisy Aircraft and the delay in getting through to the Call Centre. Not one of those words (apart from the The at the start of the sentence) should have a capital.

- To indent paragraphs, use a tab character not a series of spaces. Better still, let your software do it for you – Alt O P S and choosing the First line option in Microsoft Word.

- Use only one space, not two between sentences.

- Titles of books, songs, films and so forth are generally denoted by the use of italics rather than inverted commas or speech marks.

- Try to avoid typing words all in capitals and try to avoid using bold and underlining as such styles can become difficult to read if overused.

- Some software automatically defaults to American Letter size pages (216mm x 280mm) rather than English A4 (210mm x 297mm). Do be careful; you can check by typing Alt F U and choosing the Paper option in Microsoft Word.

Graphics

- Graphics – anything that is not text – need to be included within your manuscript. We can add these in for you but that would be an additional cost.

- There is no need to group all of your pictures together; they can be distributed throughout the book.

- If you are scanning items for use, choose a setting of 300dpi. Any lower could make the images print poorly.

- Colour does add to the cost of printing and if you have a colour image on one side of a page with just mono (black-and-white) on the other, we have to charge for two colour pages because of the nature of the printing process.

- Colour images should be in CMYK colour and not in RGB.

- It is very important that you do not use pictures, maps or images without the permission of whoever owns the copyright. You can't copy from other books, newspapers or the internet without permission.

- Images from the internet may look crisp and clear on a screen but their quality will be too poor to print well.

Cover design

- Design service is something we can offer you at extra cost and prices are on application.

- If you are supplying your own cover files, it should be in PDF format and CMYK colour rather than RGB. Please do ensure you are buying or using images of sufficiently high resolution for printing rather than low resolution internet ones).

- The design needs to include the back, spine and front as one single file and should have a minimum of 3mm 'bleed'.

- The spine width can be calculated for you please call if you require assistance with this.

Submitting your work

- When supplying your own files, we require one for the cover and a separate one for the interior pages. The text file needs to be in single page order from 1 through to the back.

- We can accept the material through the post on CD, DVD or memory stick .

- If the files are less than 8Mb we can accept them by e-mail, and if larger then you can use one of the free to use file transfer sites on the internet such as www.mailbigfile.com or www.yousendit.com

ISBN

The ISBN number is an international identification code which is unique to a particular edition of a particular title. It serves to facilitate accurate cataloguing, ordering, accounting, billing and stock control. If you purchase your ISBN numbers (minimum of 10) you will get a form with the list of numbers and you will have to assign one to your book. The UK ISBN Agency is run by Nielsen and is responsible for assigning ISBN Publisher Prefixes to publishers based in the UK or Republic of Ireland. We are therefore in the unique position of having a direct relationship with over 40,000 publishers which allows Nielsen Book's database of titles and publishers to be the most up-to-date and comprehensive.

The agency:

- Allocates ISBN Publisher Prefixes to eligible publishers based on the information provided by the publisher
- Advises publishers on the correct and proper implementation of the ISBN System
- Maintains a database of publishers and their prefixes for inclusion in the Publishers International ISBN Directory
- Encourages and promotes the use of the Bookland EAN bar code format
- Encourages and promotes the importance of the ISBN for a proper listing of titles with bibliographic agencies such as BookData
- Provides technical advice and assistance to publishers and the book trade on all aspects of ISBN usage

Contact Nielsen:

Tel: +44 (0)870 777 8712 (9am-5pm)
Fax: +44 (0)870 777 8714

Email: isbn.agency@nielsen.com

Website: www.isbn.nielsenbook.co.uk

Registered Office:
3rd Floor Midas House,
62 Goldsworth Road,
Woking,
Surrey
GU21 6LQ

About Nielsen Book

Nielsen Book has four key brands: Registration Agencies (ISBN, SAN, ISTC), BookData, BookNet and BookScan. Nielsen ISBN Agency provides publisher prefixes/ISBNs to publishers based in the UK & Ireland and offer advice on how to apply and use ISBNs. Nielsen BookData provides comprehensive, enriched and timely bibliographic data worldwide to booksellers, libraries and publishers in 110 countries around the world. Nielsen BookNet provides value-added transaction services to the industry. Nielsen BookScan provides the world's only continuous retail monitoring service for English-language books in the UK, Ireland, US, Australia, South Africa, Italy, Spain, New Zealand and Demark. The Nielsen Book group employs more than 140 staff in the UK. The company is wholly owned by The Nielsen Company, one of the world's largest publishing and information companies.

About The Nielsen Company

The Nielsen Company is the world's leading provider of marketing information, audience measurement, and business media products and services. By delivering an unmatched combination of insights, market intelligence, advanced analytical tools, and integrated marketing solutions, Nielsen provides clients with the most complete view of their consumers and their markets. Headquartered in New York, USA and Haarlem, The Netherlands, Nielsen operates in more than 100 countries with a global team dedicated to helping clients compete more effectively and discover opportunity with more clarity than ever before.

Legal deposit libraries

Legal deposit is the act of submitting published material to designated libraries and archives. In the United Kingdom the Legal Deposit Libraries Act 2003, and the Copyright and Related Rights Act 2000 (Ireland) make it obligatory for publishers and distributors in the United Kingdom and Ireland to deposit their publications within one month of publication to six libraries that collectively maintain the national published archive of the British Isles: The British Library, The Oxford Bodleian Library, The Cambridge University Library, The National Library of Scotland in Edinburgh, The Dublin Library of Trinity College, The National Library of Wales in Aberystwyth. This is a four-centuries-old tradition which will make your work available in the reading rooms, preserved for the benefit of future generations, a part of the national heritage. Your book will be recorded in the online catalogues of the libraries, all of which are accessible on the World Wide Web and provide essential research tools for generations to come.

Book Sizes

134 x 197mm (Standard)
216 x 138mm (Demy)
210 x 148mm (A5)
234 x 156mm (Royal)
249 x 177mm (Pinched Crown Quarto)
246 x 189mm (Crown Quarto)
276 x 219mm (Demy Quarto)
297 x 210mm (A4)

We can cater for almost any size, so please do ask.

We are always happy to help and offer advice so please feel free to call either Dale Burgess or one of our team to discuss your project, and requirements.

Artwork/File Guide

To ensure that we are able to deliver your job on time and exactly to your expectations we recommend that you take particular care and attention when creating and supplying your artwork. Below you will find some guidelines that will help you to artwork your design in a format that can be opened and used easily. This will also reduce or eliminate extra costs that can be incurred when files require additional manipulation in order to proceed with the print process.

File formats

Although we can accept artwork created in many different applications we have found that artwork supplied in the following formats seem to cause little or no problems. If you are supplying artwork in an application not listed below please call 01604 882588 for advice. Quark Xpress (Macintosh) | Adobe Illustrator (Macintosh or PC) | Corel Draw (PC) | PDF

Microsoft Office suite: We advise that artwork created in any Microsoft Office application should be saved as a PDF file in order to successfully send it by email. If you do not convert to a PDF the best way to send the files would be on CD or floppy disk. You should also ensure that all fonts used are embedded in the document. We do not recommend creating colour documents in the Microsoft suite for Lithographic printing as it does not handle industry standard colour models.

Scanned images within a file must be scanned in at a minimum of 600 dpi and at 100% scale. These should be saved as .jpg .tif or .eps. Ensure you use the CMYK colour model as opposed to RGB. Laser proofs should accompany all files supplied. If you are emailing a file please send a laser proof by post.

Media

Artwork should be supplied on the following media: CD ROM or Memory Stick if the files are not too large they can be emailed to sales@directpod.com

Printed proofs

Where a proof is supplied either by fax or post for your approval, it is the customer's responsibility to check the artwork. Any errors made by DPOD will be corrected and reproofed to you at our cost.
Any client amendments/author's corrections will be charged at extra cost £25 per hour.
Important note – No amendments or re-proofing will be undertaken without firstly advising the customer of additional costs and subsequently receiving written instructions from the customer.

Quick checklist

Macintosh files accepted

- Quark Xpress 3.3 and above
- Adobe Illustrator 9 and below
- Macromedia Freehand 9 and below
- Adobe PDF

Windows PC [98, XP, 2000] files accepted

- Microsoft Office 200 Applications (for digital print only)
- CorelDraw 10
- Pagemaker 6.5
- Adobe PDF

Media formats accepted

- CD ROM
- Email
- www.yousendit.com or www.mailbigfile.com

Ensure you supply

- All fonts
- All images saved as .jpg or .tif or .eps with a minimum resolution of 600dpi
- Laser proof
- Email

Direct-POD terms and conditions.
www.direct-pod.com • 01604 882588 • sales@direct-pod.com

PRINTED PAPER CASE SETUP

Hardback spine width = Paperback spine + 4mm for the boards

ACTUAL SIZE

(example shown below for A5 148mm x 210mm book)

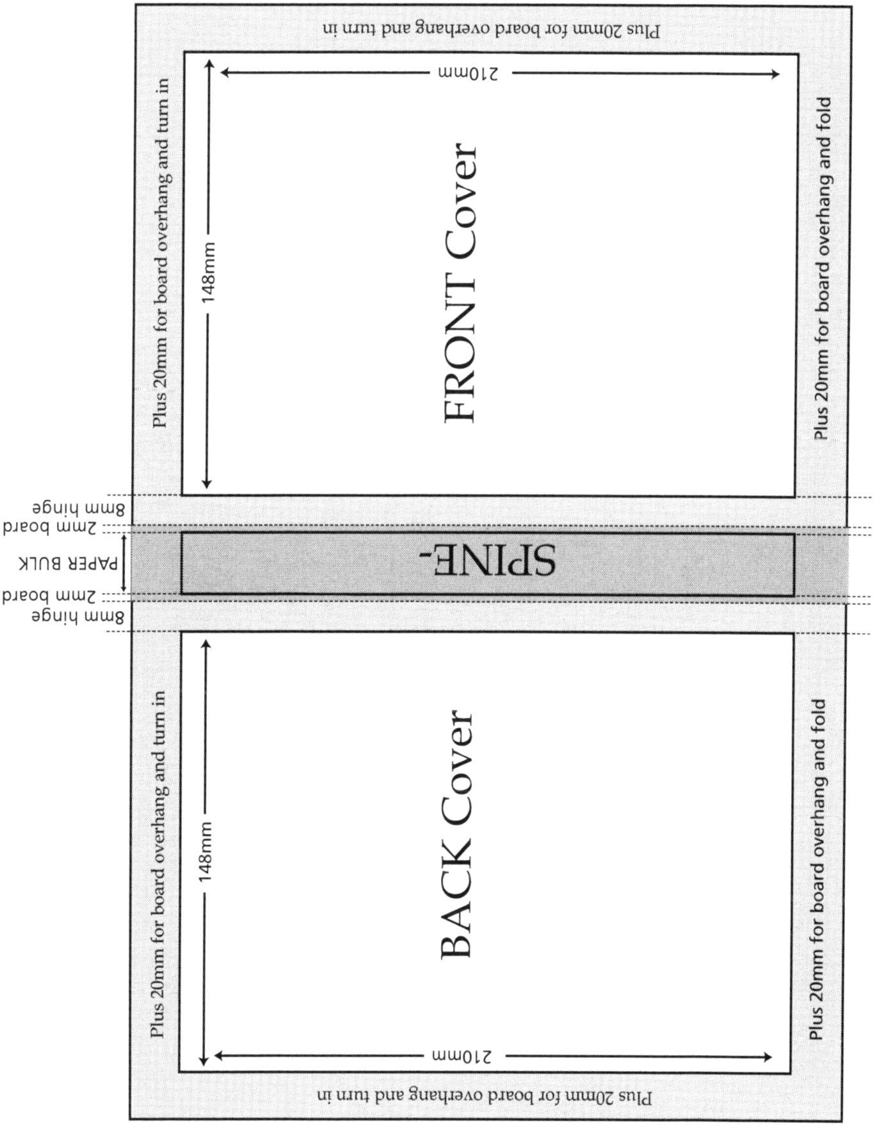

SOFTBACK COVER SETUP

ACTUAL SIZE
(example shown below for A5 148mm x 210mm paper back book)

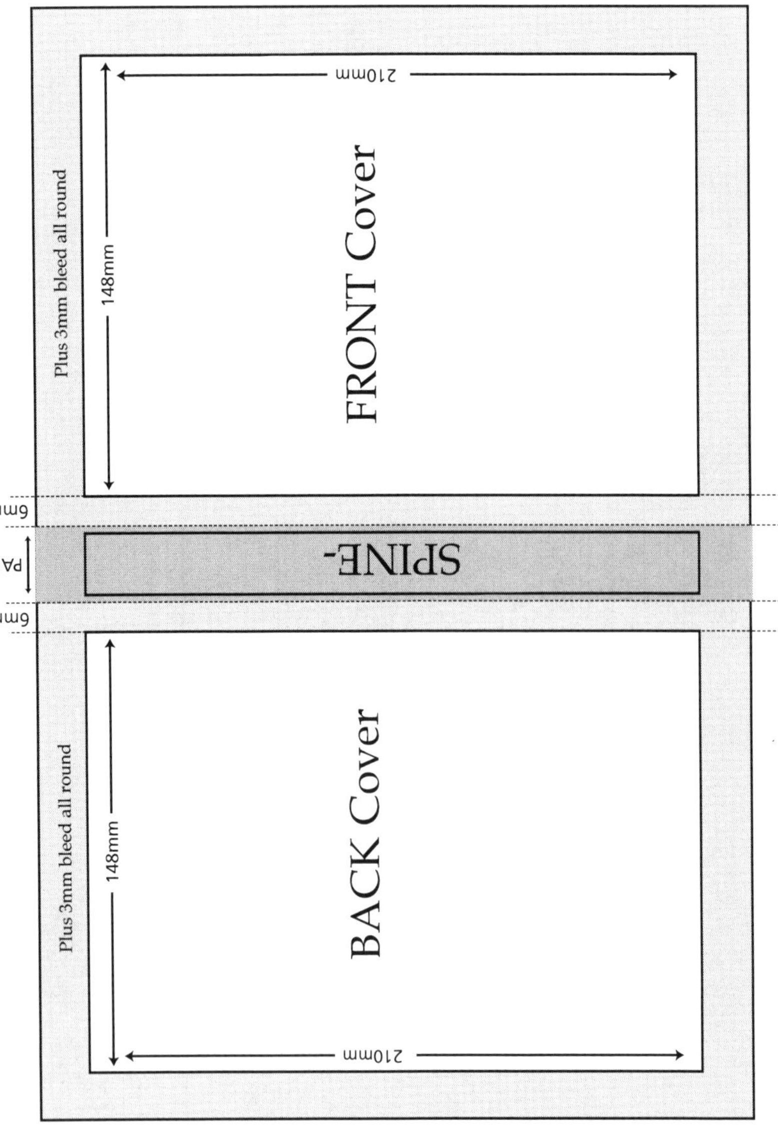

Plant list for Direct-POD

Digital b/w equipment

Oce 6250 Varioprint 1200 dpi
Maximum speed 100 impressions per minute
Paper up to 200gsm, sheet size 320 x488

Oce 2110 Varioprint 1200 dpi
Maximum speed 100 impressions per minute
Paper up to 200gsm, sheet size A3+

Oce 3155 1200 dpi
Maximum speed 55 impressions per minute
Paper up to 160gsm, sheet size A3

Our digital colour equipment

Oce CS650 1200 dpi
Maximum speed 65 impressions per minute
Paper up to 300gsm, sheet size SRA3 (320 x 450)

Oce CPS 900 1200 dpi
Maximum speed 30 impressions per minute
Paper up to 300gsm, sheet size 305 x 457mm

Lithographic printing equipment

Komori Lithrone 5 colour
Maximum speed 15000 impressions per hour, paper up to maximum size SRA2.

Heidelberg GTO 52 Including numbering, perfing etc in line, paper up to maximum size SRA3.

Finishing

To complement our print services we provide a complete range of finishing including, Perfect binding, wiro and spiral binding, drilling, cutting, folding, creasing, laminating, encapsulating, shrink wrapping, collating and labelling.

Mailing and fulfilment is a rapidly expanding part of our business. From receipt of your database we can mailsort, print personalised letters and other promotional material, supply envelopes -printed or plain, enclose and prepare your items for Post Office or courier collection.

We provide guaranteed next or same day despatch throughout the UK and Europe as requested.

Pricing

POD Prices for 1-25 copies

Our POD 1-25 copies is based on a contracted agreement, and a volume of titles and orders. A copy of our contract is available on request.

The basic pricing is as follows:

Annual Charge per title	£10.00
Set-up cost per title one off	£17.50
Printing and perfect binding costs per copy:	
Cover bind and gloss lamination A5 and below	£0.70
Cover bind and gloss lamination A5 and above	£1.00
Text mono printing A5 and below	£0.01
Text mono printing A5 and above	£0.02
Extra costs:	
Scan and recreate text and cover	£50.00
Perfect Bound Proof	£25.00
Loose leaf proof	£15.00

All prices subject to sight of files and or hard copy, and full specification of the title(s). Terms and conditions available on request.

A5 Books

DIRECT-POD
THE PRINT ON DEMAND COMPANY

Saxon Fields
Old Harborough Road
Brixworth
Northants
NN6 9BX

Telephone: 01604 882 588
Fascimile: 01604 882 595

e-mail: dale@direct-pod.com

www.direct-pod.com

Setup	35	15	15	15	15	15	15	15	15	15	15
Delivery	8	13	18	23	28	33	38	43	48	53	58
Quantity	1	50	100	150	200	250	300	350	400	450	500
No. Pages											
496	65	253	427	627	813	970	1,152	1,332	1,510	1,685	1,858
480	57	235	419	614	797	950	1,129	1,305	1,479	1,650	1,820
464	57	231	410	602	780	930	1,104	1,277	1,447	1,615	1,780
448	56	226	401	588	763	909	1,080	1,248	1,414	1,578	1,740
432	56	221	392	575	745	888	1,054	1,218	1,381	1,541	1,699
416	56	216	383	561	727	866	1,028	1,189	1,347	1,503	1,657
400	56	212	374	547	709	844	1,002	1,158	1,312	1,464	1,614
384	56	206	364	533	690	822	975	1,127	1,277	1,424	1,570
368	56	201	355	518	671	799	948	1,095	1,240	1,384	1,526
352	56	196	345	503	651	775	920	1,063	1,204	1,343	1,480
336	56	191	334	488	631	752	891	1,030	1,166	1,301	1,434
320	56	185	324	473	611	727	862	996	1,128	1,258	1,387
304	56	180	314	457	590	703	833	962	1,089	1,215	1,339
288	56	174	303	441	569	677	803	927	1,050	1,171	1,290
272	56	168	292	425	548	652	772	892	1,009	1,126	1,240
256	56	163	281	408	526	626	741	856	968	1,080	1,190
240	56	157	270	391	504	599	710	819	927	1,033	1,138
224	56	151	259	374	482	572	678	782	884	986	1,086
208	56	145	247	357	459	545	645	744	841	938	1,033
192	56	138	235	339	436	517	612	705	798	889	979
176	55	132	223	321	412	489	578	666	753	839	924
160	55	126	211	303	389	460	544	626	708	789	868
144	55	119	199	285	364	431	509	586	662	737	812
128	55	113	186	266	340	401	474	545	616	685	754
112	55	106	173	247	315	371	438	504	569	633	696
96	55	99	161	227	289	341	402	461	521	579	637
80	55	92	147	208	263	310	365	419	472	525	577
64	55	85	134	188	237	279	327	375	423	470	516
48	55	78	121	168	211	247	289	331	373	414	454
32	55	71	107	147	184	215	251	287	322	357	392

A4 Books

DIRECT-POD
THE PRINT ON DEMAND COMPANY

Saxon Fields
Old Harborough Road
Brixworth
Northants
NN6 9BX

Telephone: 01604 882 588
Fascimile: 01604 882 595

e-mail: dale@direct-pod.com

www.direct-pod.com

Setup	35	15	15	15	15	15	15	15	15	15	15
Delivery	10	15	20	25	30	35	40	45	50	55	60
Quantity	1	50	100	150	200	250	300	350	400	450	500
No. Pages											
496	69	436	770	1,142	1,488	1,780	2,119	2,453	2,783	3,109	3,431
480	58	412	753	1,117	1,455	1,741	2,072	2,398	2,721	3,040	3,354
464	58	403	736	1,091	1,422	1,700	2,023	2,342	2,658	2,969	3,276
448	58	394	719	1,065	1,387	1,659	1,974	2,285	2,593	2,896	3,196
432	58	384	701	1,038	1,352	1,617	1,924	2,227	2,527	2,822	3,114
416	58	375	683	1,011	1,316	1,574	1,873	2,168	2,459	2,747	3,031
400	58	365	664	983	1,280	1,530	1,820	2,107	2,390	2,670	2,946
384	58	355	645	955	1,243	1,485	1,767	2,045	2,320	2,591	2 859
368	58	345	626	926	1,205	1,440	1,713	1,982	2,248	2,511	2,770
352	58	334	606	896	1,166	1,393	1,657	1,918	2,175	2,429	2,680
336	58	324	586	866	1,126	1,346	1,601	1,852	2,101	2,346	2,588
320	57	313	566	835	1,086	1,298	1,543	1,785	2,025	2,261	2,494
304	57	302	545	804	1,045	1,249	1,485	1,717	1,948	2,175	2,399
288	57	291	524	772	1,004	1,199	1,425	1,648	1,869	2,087	2,302
272	57	279	502	740	961	1,148	1,364	1,578	1,789	1,998	2,203
256	57	268	480	707	918	1,096	1,302	1,506	1,708	1,907	2,103
240	57	256	458	674	874	1,043	1,240	1,434	1,625	1,814	2,001
224	57	244	435	640	830	990	1,176	1,360	1,541	1,720	1,897
208	57	232	412	605	785	936	1,111	1,284	1,455	1,624	1,791
192	56	220	389	570	739	880	1,045	1,208	1,369	1,527	1,684
176	56	207	365	534	692	824	978	1,130	1,280	1,429	1,575
160	56	194	341	498	644	767	910	1,051	1,191	1.328	1,464
144	56	181	317	461	596	709	841	971	1,100	1,227	1,352
128	56	168	292	424	547	650	771	890	1,007	1,123	1,238
112	56	155	266	386	497	591	700	807	914	1,019	1,122
96	56	141	241	348	447	530	628	724	818	912	1,005
80	55	128	215	309	396	469	554	639	722	804	885
64	55	114	188	269	344	407	480	553	624	695	765
48	55	100	162	229	291	343	405	465	525	584	642
32	55	85	135	188	238	279	328	377	424	471	518

Book Printing Prices for 25-250 copies

Set Up Costs

Set up charge	£35.00p
Scanning per page from hard copy supplied	£00.10p
Work to scanned pages with tints and halftones on charged extra by the hour @ £25.00	
Scanning cover from hard copy supplied	£50.00p
Loose leaf proof	£15.00p
Perfect bound proof	£25.00p

Print Costs black and white	**Price per page**
Standard size 132mmx197mm	1p
Demy size	1p
A5 size	1.2p
Royal size 234x156mm	1.2p
A4 size	1.5p
Colour page inserts in a mono book	15p

If the colour pages exceed 20% of the total pages the quote will be based on a full colour book.

Print Costs Colour	**Price per page**
Standard size 134mmx197mm	3.5p
Demy size	3.5p
A5 size	3.5p
Royal size 234x156mm	6p
A4 size	7p

Finishing costs	**Priced per unit**
Perfect bound with gloss or matt laminated cover	95p
Case bound with gloss or matt laminated PPC upto 100 copies	£6.00
Case bound with gloss or matt laminated PPC upto 100-250	£4.50
Wire-O Binding with gloss or matt laminated cover	£1.45
Saddle Stitching with gloss or matt laminated cover	95p

Additional charges

80gsm Opaque	No charge
100gsm Opaque	6% extra
90gsm Volume 18 Bookwove	10% extra
120gsm Gloss	Price on application
Other stock weights	Price on application
Delivery charges	£8.00 plus 10p per book

We have a minimum charge of £50.00 per order.
Our standard terms and conditions are at the back of this book.
Prices are correct at time of printing this sample, but please note they are subject to change.

Further information available at www.direct-pod.com

Book Printing Prices for 250 copies plus

We are set up to offer volume runs and pricing, for 250 copies plus. Offering a range of various sizes, and with or without colour sections, also in hardback or soft back. Currently prices for these run lengths are price on application so please call today and speak to one of our customer services team who will be happy to assist with your enquiry.

Because the latest technology helps up to keep costs down and quality up we can offer these larger runs at a cheaper rate than litho and in a quicker scheduled time from order placement to delivery.

We use the latest Oce technology which offers you a 1200 dpi print option for high quality halftones and tints.

For colour insert prices please see the previous pages of our book where it states the colour page insertion and printing cost for you.

Example 1 size upto A5 max:
From files supplied, loose proof and print
Text 160 pages black only on 80gsm Opaque
Cover 4/0 on 280gsm Trucard and gloss laminated
Collated and perfect bound finish (softback)
Boxed and delivered to one UK address
250 copies £523.00

Example 2 size upto A5 max:
From files supplied, loose proof and print
Text 112 pages black only on 80gsm Opaque
Cover 4/0 on 280gsm Trucard and gloss laminated
Collated and perfect bound finish (softback)
Boxed and delivered to one UK address
1000 copies £1417.00

DIRECT-POD

THE PRINT ON DEMAND COMPANY

I'll have a 'P' Bob...

Océ

Digital Book Production

Benefit from new value chains

The Book Market – an Océ Perspective

- Changing reading habits
- Altered buying behaviour
- Shorter life span of knowledge
- Wide variety of titles
- Process optimization
- Structural change

Benefit from new value chains

Océ Digital Book Production
The Book Market – an Océ Perspective

In 2006, the global book market achieved sales worth 100 billion euros. Looking at Europe, Germany is a particularly strong market, publishing over 94,000 new titles and producing 970 million books in 2006. With worldwide annual production of about 900,000 titles, this means one in ten books is published in Germany. Other significant players on the European new-title leaderboard are France and Spain, with annual productions of 68,000 and 63,000 titles respectively.

However, the runaway leader is the United Kingdom, which also supplies many other countries with English books. The UK published in excess of 160,000 new titles in 2006, which marked yet another significant increase over the previous year's 130,000.

Changing reading habits

Information alone is not knowledge. What matters is how people engage with and respond to facts, arguments and information, process input, and transform it into memory. Reading is still an effective way of doing this. And it works with unrivalled efficiency. Yet consumer reading patterns are changing. In 2000, the German "Stiftung Lesen" – a foundation to promote reading – surveyed a thousand people aged over 14 on their reading habits. As well as the general survey, in-depth interviews were conducted to determine how different types of readers engage with print media and the Internet, providing greater insights into reader profiles.

A comparison between these results and a 1992 study reveals a sharp drop in the numbers of people who read books daily. Only 6 per cent dip into a book every day; eight years ago the figure was 16 per cent. Once embedded in daily routines, book-reading is now becoming more an ad-hoc leisure pursuit. People are nowadays more likely to pick up a book when they have the time – at weekends, for instance – and are in a relaxed frame of mind. The idea of reading something thoroughly from start to finish is fast becoming obsolete, especially among young people.

Nearly one in three respondents under 19 admitted: "I sometimes skim-read the pages and pick out the bits that look most interesting". In 1992, only one in ten was a skim-reader. The same behavior is observable – albeit less pronounced – across all age groups. The altered behavior patterns docu-mented by study were characterized by Bodo Franzmann, a researcher at the foundation, as "an adaptation of reading strategies to the relentless information overload of our media society". In his view, these enduring changes to our reading culture are set to continue.

Franzmann points out that reading abilities are deteriorating. "Instead of focusing on the text, many readers are now developing the tendency to zap through pages in much the same way as they would zap TV channels." Franzmann does not conclude that the book in its classic form is under threat from the Internet or e-book, but emphasizes that publishers need to respond to these changing patterns. Many people are reading less, he says, but this reading is far more targeted. And this means highly specialized publications with small target audiences are becoming increasingly important.

Altered buying behaviour

Source:
Hugendubel Munich
Fünf Höfe, Theatinerstraße 11

However, it is not only reading behavior which has changed. There are also differences in how books are offered for sale, and how they are purchased. When they walk into a bookshop, consumers are getting used to finding piles of special offer books at the entrance and a huge choice of instock titles to browse through once they're in the shop. Adapting their presentation and distribution strategies, traditional physical bookstores – long thought to be dependent on publishers – have evolved into players with genuine market power. Large retailers and chains, supermarkets selling books, Internet bookshops such as Amazon all demonstrate the diversity of approaches.

And literature is also becoming increasingly international. Approximately one third of fiction and non-fiction portfolios in Germany is now translated from another language. As might be expected, three quarters of translations are from English, but the more "minor" literary languages such as Swedish, Dutch or Polish also account for numerous titles. Horizons are expanding, certainly, but this makes it all the more difficult for readers to navigate the world of literature. Spur-of-the-moment purchases are one obvious way to respond; either that, or consumers buy books which have increasingly derive their content from media such as television, the music scene, computers or the cinema, and also rely on exposure through these media for their publicity. From classics sold on the back of film and series adaptations, to books by TV stars, this happens with virtually every subject. Even independent literature, if discussed (or rather treated) on TV talk shows, can take on the flavor of a TV follow-product. In scenarios such as this, the book's intrinsic importance has indisputably declined. The book itself is successful because the alleged competition – television and Internet – have given it plenty of exposure.

The direction is clear; and there's a way to deal with the challenge: through printing on demand. Regardless of the category – from short-run special-interest literature to speculative bestseller editions – more and more books have to be printed while they're "hot". It's increasingly difficult for publishers to guesstimate the quantities they are going to sell. There is also an ever increasing risk of ending up with piles of unsaleable inventory. Here too, digital printing can build the bridge to combine profitability with superfast response times.

Shorter life span of knowledge

Our information overload generates its own new information. The fields of science, research, engineering, development, also descriptive and reporting disciplines such as journalism and non-fiction, are accelerating time-to-market while at the same time cutting the life span of explicit knowledge. What is true and correct today might be known, seen and presented differently tomorrow. Knowledge is inflationary. Depending on the field, the half-life of knowledge is now no more than three, perhaps five years. There are far fewer fundamental insights and theories around today which can stake a genuine claim to any degree of permanence.

Sometimes, as in information technology and other hightech industries, facts, skills, knowledge and philosophies change in the span of months. In the IT sector, six months is an entire generation. Books are barely off the press before they are outdated.

By the time they leave university, new graduates know that what they learnt at the start of their courses is no longer up to date – not just in terms of minor detail but also regarding the real "meat" of the subject. It might sound cynical, but the axiom "never trust information whose age you don't know" is basic pragmatism. Yet it's not only documented knowledge where the rate of change is accelerating. The pace is also picking up in aspects of everyday life such as fashion and lifestyle. How long a thing can claim to be new, current or modern is of elementary importance, above all for printed media.

The life of content or entire titles is reduced to just a few months, impelling publishers to change content and print updated editions within months of publication. A technology that can deal with this challenge – as digital printing can – is therefore highly welcome.

Wide variety of titles

Predicting the behavior of book buyers is a mounting challenge for publishers. Everything is in flux: demands are increasingly differentiated, interests, purchasing power and readiness to purchase are all changing. Customers no longer buy just what they are offered. They compare and contrast, weigh up decisions and select critically. No wonder, then, that the publishing industry feels compelled to respond to this development with a deluge of new titles every year. But each new title runs the risk of being a flop.

Remaindering – selling overstock – is a normal part of the industry. In fact, right from the start, books are printed in high enough volumes to sell off at knockdown prices as "remainder copies". The drive to present the image of a booming industry is exacerbating the dilemma of overproduction. Every year sees a fresh record number of new and re-issued titles. In 1990 about 61,000 books were published in Germany; 15 years later we are seeing close on 90,000 titles per year – a rise of over 45 per cent.

By way of contrast, average run lengths are falling. A 3,000-unit run was long considered the lowest limit. Any lower, and sales would most likely not cover production costs. Thanks to digital printing, books with a run length of 300 copies, sold to interested parties, are often more profitable than 30,000 copies sold via the remainder bins. The publishing risk is increasing.

Since 2000, industry sales volume has lagged behind the rising number of titles, and the proliferation of titles is a profitability problem for the industry. More titles are produced, yet unit volumes are staying the same – meaning the lengths of individual runs are getting shorter. The traditional production process, measured against the benchmarks of flexibility, updateability and short runs, increasingly fails to deliver the required profitability. The only alternative, the expedient that really works, is printing on demand: digital printing in all its shapes and forms.

Process optimization

Essentially there are three characteristics that make digital printing so appealing for publishing production:

- Digital printing offers significant cost advantages over conventional methods for printing short runs (under 2,000 units). This is due to the relatively low fixed costs.
- Digital printing offers unparalleled flexibility. It permits last-minute changes and content updates because there is no static information substrate (printing plates).
- Digital printing permits relatively short production times. By turning out ready-collated copies, it accelerates finishing and enables just-in-time delivery. Even single copies can now be printed economically (in digital printing, as in other industries, "mass customization" is driving trends).

The intelligent combination of these properties is the enabler for on-demand production of books in exactly the quantities required by the market. Yet publishers have so far been hesitant to try out the opportunities of digital printing. This low buy-in is down to a few small issues - but as the saying goes, the devil is in the detail. The problem has mainly concerned the post-print finishing and binding processes.

Although high quality printing of content on typical book paper is now relatively easy to implement, it has proved difficult - if not impossible - to produce single hardback covers. A solution has now arrived for this challenge also, resolving the problem of this work step. However, seen objectively, digital print providers might like to take a self-critical look at their own attitudes. With starry-eyed accolades, often using inappropriate examples or execution, providers have been carried away by their enthusiasm, hyping up solutions in a manner that was simply out of synch with the ideas of a highly vocal and critical phalanx of publishers and publishing directors. But these teething problems have been overcome.

The awareness and insights are already in place as to what publishers really need - and the digital printing alternative can demonstrate solutions along the entire production process chain. From years of experience, we at Océ know where the strengths and weaknesses of each step in the process lie, and how conventional production (offset) and digital printing can best be teamed together. Océ can therefore advise publishers on business cases for transitioning their portfolios to digital.

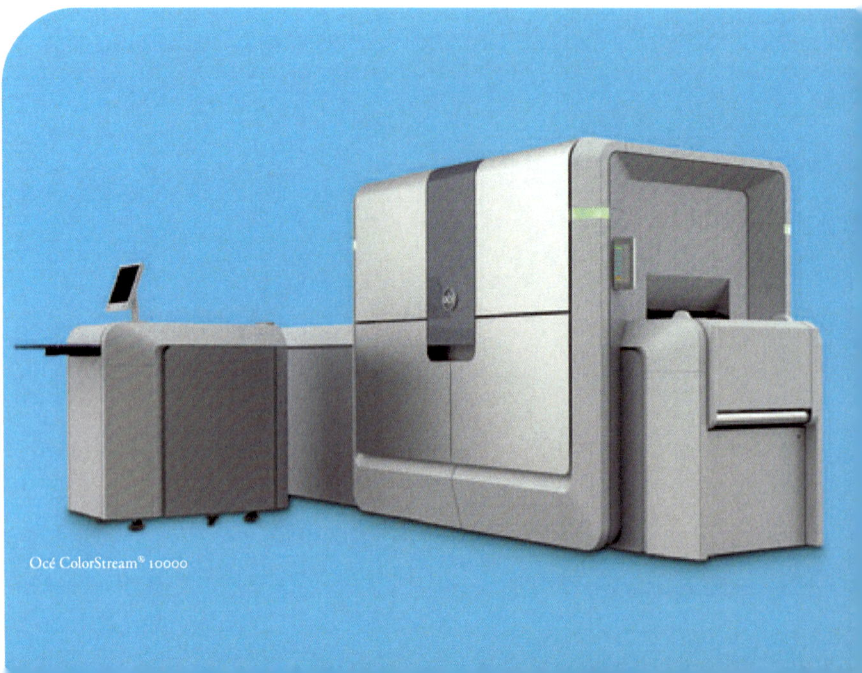

Océ ColorStream® 10000

Yet digital book production requires more than technical knowledge. Publishers expect complete, lucid proposals – it might sound trite, but there is a genuine shortfall on this. From the publishers' viewpoint, the vista of digital printing opportunities and choices remains highly bewildering. To work out how to deploy technology most profitably for any given task or production step, it is essential to analyze the entire value chain. Just comparing unit costs fails to do justice to the complexity. The comparison must take in the entire process from author to reader. The publishers manage this process, but do not execute all the steps themselves. Beyond the cost of production, publishers have a range of other costs: storage costs and the cost of capital, for instance. To identify which production method is the most profitable for the title in question, other process and cost chains have to be calculated:

- Printing costs
- Warehousing and distribution costs
- Pulping / returns
- Sales period
- Cost of capital
- Updates
- Planned run length

Océ offers publishers a concept by which they can produce each of their titles, and keep them available, using the most economical method. In Germany, 300,000 unsaleable books are pulped every week – that's approximately 15.6 million books per year. These books cost publishers in excess of one million euros annually, running up costs for inventory management and logistics as well as in production. The book market offers only slim chances of increasing profits by sales growth – so the obvious route is to raise profitability by optimizing costs, i.e. by raising returns per title or book.

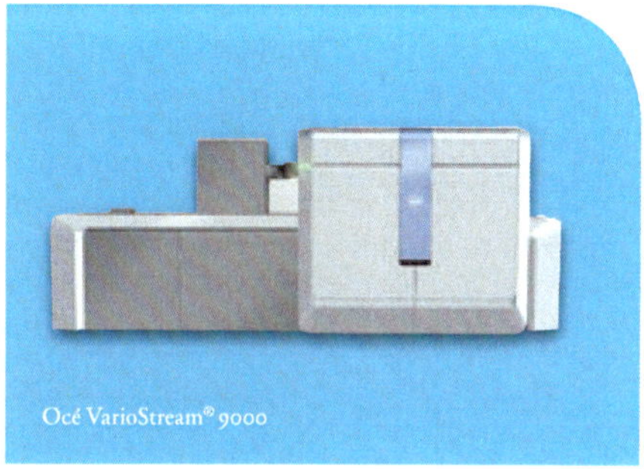

Océ VarioStream® 9000

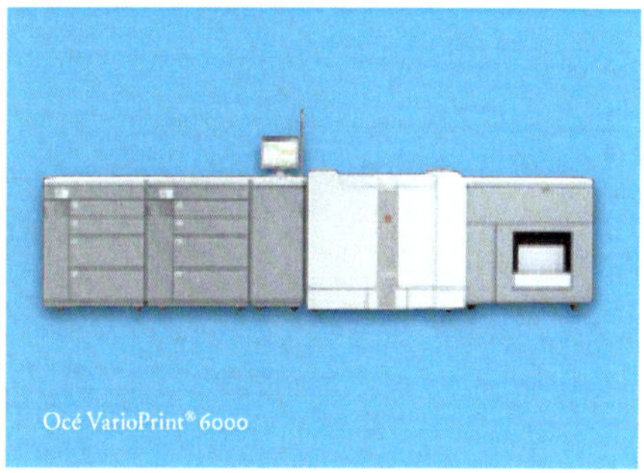

Océ VarioPrint® 6000

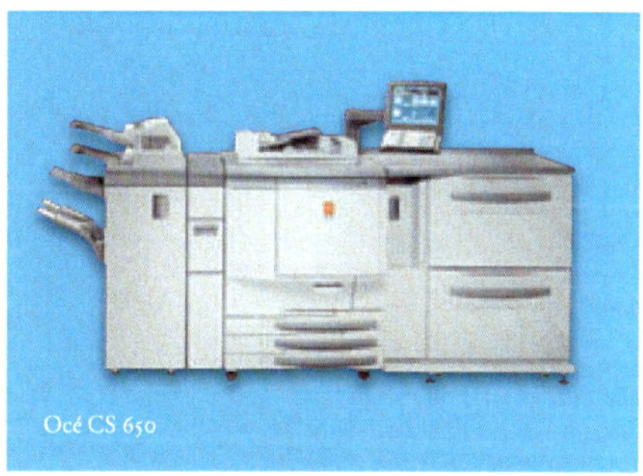

Océ CS 650

Océ JetStream™

Structural change

Growth in publishing revenues is achievable only through an optimized blend of marketing and production activities that also embrace e-commerce.

More and more books are self-published by businesses and organizations or private individuals who take charge of their own marketing and production. Marketing aspects, and direct contact with and access to customers, play a crucial role. The winners are the book projects being managed and controlled directly by consumers. These are projects which use the possibilities of digital communications technology as their value driver. The best example: photo books, created on print portals and printed fully digitally. "Vanity" publishing – where authors themselves take charge of book production and marketing – uses much the same principle.

Digital printing-based services can also be used by publishers for more effective management of backlist titles. One excellent example is the service offering of Lightning Source, which has locations in the USA and UK and reprints approximately one million books per month on Océ printing systems for some 450 publishers. The average run length is 1.8 books, with delivery within 12 or 24 hours of order.

Publishers account for a steadily dwindling percentage of worldwide book production volumes. Their share has already dipped below 40 per cent. More and more books are produced and marketed by third parties using non-traditional methods. Publishers who can occupy appealing niche themes with innovative products will retain a stable base if they consistently leverage the opportunities of digital printing and the associated new manufacturing concepts. The combination of web and digital printing also creates novel opportunities to take books beyond their role as static products and make them useable as tools in a culture of communication which is end-customer-driven. However, marketing – for series titles, individual titles and publishing companies – is essential to build trust for the publishing brand, communicate to customers what the brand stands for, and help to navigate them through their purchasing decisions.

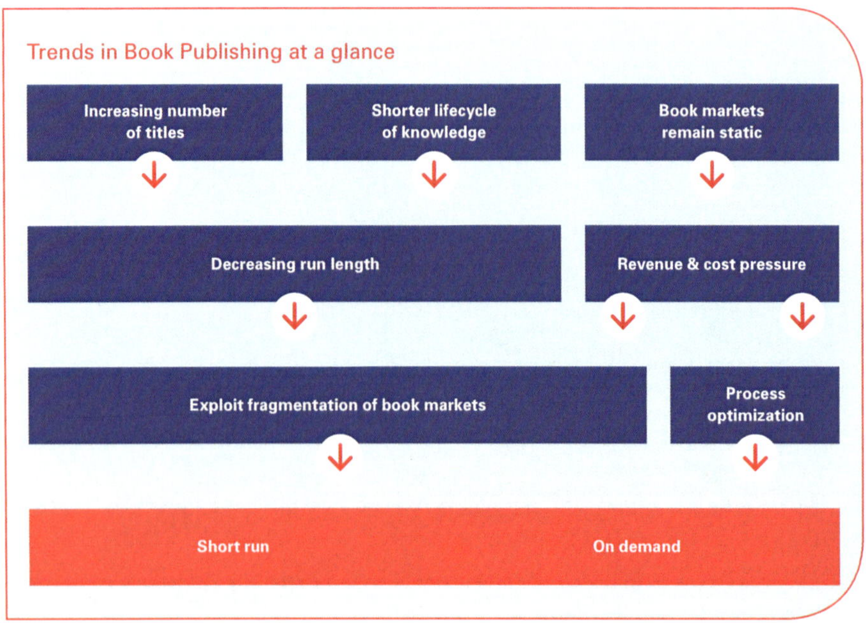

"The supply chain in the book industry is extremely inefficient. Our vision is to collapse the supply chain as much as possible and apply the principles that have worked in other industries to get the right books out at the right time in the right quantity.
... With our Océ equipment we have achieved industry-leading quality with halftones and solids. Direct-POD can now offer the market the highest level of print on demand quality."

Beyond the Ordinary

océ

Printing for Professionals

Océ helps the people who make our world. Companies everywhere use Océ technical documentation systems in manufacturing, architecture, engineering and construction. Each week, high speed Océ printing systems produce millions of transaction documents such as bank statements and utility bills. And in offices around the world, people use Océ professional document systems to keep the wheels of business and government turning. Océ is also at work in publishing on demand, newspaper production, document management outsourcing and wide format colour for spectacular display graphics. It all helps our professional customers go 'Beyond the Ordinary' in printing and document management.

War & Peas
(or very interesting facts you may not know about Books & Print)

Would it surprise you to learn that a printer invented the ghost?

Well, when I say "ghost" I don't actually mean the spooky shimmery beings associated with the deceased, but rather the word "ghost".

And to be more precise the word "ghost" spelt "ghost" rather than the earlier Middle English "gost"

It is said that the spelling "ghost" with the silent "h" was brought to English by Caxton as a result of the spelling habits he picked up in Bruges where he had been living. The word "spook" also comes from the Dutch, but that's another matter entirely.

Caxton was the 15th century English merchant, diplomat, member of the household of the Duchess of Burgundy, dealer in cloth, writer, translator, and bookseller who introduced the printing press to England.

He printed the first ever book in English (Recuyell of the Historyes of Troye), before returning from Bruges to produce the first book printed in England: The Canterbury Tales.

But like all publishers (and I here I speak from the heart) Caxton was not universally loved. This wasn't that he was smelly, or particularly odd looking, but rather because the upper classes worried that if ever the poor got to read they would finally wake up to the fact that they were, well, poor.

However being the first man to print in English gave Caxton a problem. Then, as now, the language was changing and he was the first one to have to face the issue of standardising the language and deciding what words meant and how they were spelled.

Caxton produced one book at a time, and that's exactly what we do with digital printing (although of course you can have 50 or 250 or more if you want).

And as we work, the language slips and slides around us – which is exactly what Caxton felt about English.

So too round of my discourse here are four recently introduced words: Hardscape, Radge, Retronym, Crunk. All are recognised by the OED.

If you can provide us with any book to print which incorporates the use of any three of those words in a proper and meaningful way, we will give you a 10% discount.

But really, you don't need it. Our prices are very reasonable anyway.

And you need only print one at a time.

In the strange world of printing, nothing is ever quite what it seems.

Well, that's not right because paper is generally called, well, paper.

But what about the hardback book? The sort of book you expect to dominate libraries. The sort of book you would probably prefer to give away as a birthday present (on the grounds that a paperback always looks a bit cheapskate as a gift.) It is a book made with stiff outer covers and usually covered with cloth, vinyl or leather. It can also come with a dust jacket.

Sadly, in the odd world of print, the hardback is often not called a hardback at all. It is called a casebound book.

Worse is to come. The binding of a hardback involves things called "squares" which aren't, "the endpaper" which isn't, the "first and last signatures" which are not signed, and "tipping" which has nothing to do with slipping a fiver to the waiter.

The point about casebound books is that they have the air of quality and importance. One may know nothing about the content, but one perceives it to be of greater importance because this is a hardback. It is built to last.

Of course, hardback books cost more to print than paperbacks, but then on the other hand they sell for more than paperbacks, and generally speaking there is a greater opportunity for a bigger margin.

And, as with paperbacks and spiral bound books, there is every opportunity to print small numbers on the digital press – so your risk in terms of volume printing is much diminished.

Anyway, this is all apropos the fact that this is **Casebound Fortnight**. Now it is true that you might not find anything about this event on the internet (largely because I just made it up) but if you get a quote from us for anything casebound in the next fortnight then I will personally give you a 10% discount.

Now I must admit I haven't told the boss about this yet, but he's a jolly nice fellow, so I am sure it will be ok. And besides, hardbacks are good for you. They make you feel, sort of, good. I was going to tell him… but then…

But then my mind often wanders – as indeed my body has. You may have come across me before with another company, but I would like to say that the name, address and phone number below is right, and this is where I am. Which is different from where I was. If you see what I mean.

Just how fancy do you want your book to be?

A really smart large format paperback with a laminated coloured cover can look excellent, professional, and the sort of thing you might invest in when browsing Waterstones.

As a publisher having such a book printed might set you back £4 a copy or so – depending of course on how many pages you want.

But, I am sorry to report, not everyone is that readily pleased.

Take the case of Francis Sangorski and his mate George Sutcliffe. These two fellows met when they were doing bookbinding evening classes run by the London County Council (the organisation that ran London before the days of the GLC.)

Now you would think that meeting a like-minded chap at book binding evening classes is a pretty harmless thing to do. You might toddle off to the pub after the session, and discuss the finer points of double wire binding and the twin loop technique over a Guinness or two.

Indeed this is how I imagine the pair, gently eking out the £20 a year bursary they had each won to study the trade.

But somehow the earnings went to their head, and in no time at all they had left the world of £20 a year bursaries behind as they established their own company at the start of the 20th century. And then they reinvented jewelled bindings.

Of course they didn't stick jewels on any old book. I mean, a bejewelled edition of the British Rail national timetable might have been of interest, but would have seemed a bit incongrous.

Instead they bound an edition of the Rubaiyat of Omar Khayyam. Unfortunately, while their bookbinding skills were undoubtedly of the highest, they had forgotten to attend the soothsaying classes that ran next door, and so they sent their book to America on the Titanic in 1912, which in hindsight was not the best of ideas.

It is also a fact that the binding of the book took more attention than the text itself – which was perhaps wise because the translation that the binders chose to embed in jewels and gold was the Fitzgerald translation, a mishmash of text, some of which can be traced back to the original but some of which is pure invention, and not much of which is in the right order.

So maybe the old phrase about telling books by covers has some merit in it – and perhaps one might also add, "If binding your book in jewels, do look at the name of the boat that carries it."

I deeply regret to say that our jewelled bookbinding division is closed today, but call me if you would like to talk through any other style or approach.

My mind often wanders – as indeed my body has. You may have come across me before with another company, but I would like to say that the name, address and phone number below is right, and that is where I am. Which is different from where I was. If you see what I mean.

What's the first thing?

When asked, "what is the first thing you think of when planning a book?" most publishers answer, "what the book's about".

Which seems fair enough… except, it isn't.

There are two other questions that should be asked long before the issue of the contents of the book are considered.

First there is the issue of "who is going to buy this book and how am I going to reach them?"

Now I appreciate that logically to answer this question you need to know what the book is about – but my point here is that you only need to know that in general terms.

Let's take an example. A publisher decides to publish on the subject of steam railway engines. That's the general area of interest.

At that point he or she needs to be thinking about how potential buyers of books on railway engines are going to be attracted to such books. Only when that has been done does the question of "what exactly is this specific book about?" come to the fore.

And even then there is still the second question to answer before detailed work on the text begins.

That second question is, "what is the size of the book going to be?"

This may seem an odd point to raise before we get going in terms of the book's contents, but, in fact, facing this question before the text is set out can save you an absolute fortune.

Here's why. Our digital book printing presses print on paper measuring 320x488mm. So it is clear at once that if you choose to publish a paperback book measuring 210x297mm (A4) you will get two printed pages per sheet. If you publish a book measuring 148x210mm (A5) you will get 4 pages per sheet. We have printed some books of late at 150x150mm with 6 pages per sheet – obviously costing even less.

So my big suggestion of the day is this.

If you are about to start the long publishing journey for a new title, spend a little time considering the marketing and a little time considering the size.

A picture is not always worth one thousand words

"A picture is worth one thousand words" is one of those old adages I could really do without.

It is not that I don't like pictures or that I don't want pictures in the books we print. But it is just that for some people the saying seems to suggest that, "every book must include pictures".

The extension of this view is that ANY picture is better than none – which is not what the adage says at all – and that is certainly not true.

The "thousand words" phrase was introduced by Fred R. Barnard in the advertising trade magazine Printers' Ink. He wanted to advertise his ability to design visuals to appear on the side of trams – so he made up the phrase.

His business obviously didn't really flourish with that advert because he came back to the same theme a few months later and this time he upped the ante with, "One Picture is Worth Ten Thousand Words." He called the phrase "a Chinese proverb" and a little later wrote that it was attributed to Confucius – which isn't actually true (although others have repeated the assertion since).

Perhaps a better phrase for us to use would be that of Bonaparte who said, "Un bon croquis vaut mieux qu'un long discours" or "A good sketch is better than a long speech".

The benefit of this phrase is the inclusion of "good" – for it reminds us that pictures are not automatically of value – only pictures which expand or quickly summarise what the text says are of true merit within a book.

In modern digital printing you can include black and white pictures at no extra charge. But colour pictures are a different matter simply because they use far more ink and require a different printing process than a monochrome page.

Since the black and white pictures don't cost any extra to include in a book they can go on the page where they are relevant. Colour pictures however should be positioned on the front and back of an individual leaf (two pages that back each other) to keep the costs down (that is to say, if you have a colour picture on page 21 you can put one on page 22 as this is more cost effective – but the same is not true of pages 22 and 23 which will be opposite each other).

For a book up to A5 size a double colour page (as in the above example) will cost about 5p per copy per printed page. For sizes between A5 and A4 the price is about 8p per printed page per copy.

If you would like to talk through any sayings that Confucius never said or indeed the use of pictures in your book, do give me a call on 01604 882588.

The book is printed – what then?

While we sweat blood and move heaven and earth to give you exactly and precisely the book you want, even we have to admit that, once the printing is done, a second major task lies ahead – selling the copies of the book.

There are two quite separate issues in arranging the selling of books that you publish – the preparation stage, and the marketing stage. I look at them below…

The Preparation Stage

This involves getting the book ready so that it can be sold. It includes such actions as adding an ISBN and bar code to the book, getting the book onto the national database of book titles, and ensuring that your legal obligations are met by sending copies to the national libraries. (This last stage is worth taking seriously since this also means that your book shows up in the British National Bibliography – a major reference source.)

The marketing stage

There are many different ways of marketing a book – and although the ideas set out below are used by most publishers, they are just examples of what you can try to do. It is possible to go through other routes.

Firstly, you might want Amazon to list and stock your book. Many people use Amazon as a buying centre, so getting the book listed there is helpful. But listing is only part of the arrangement – having the book stocked by Amazon is a good move too.

Secondly, you might want to talk with outlets likely to sell your book. For example, if you have published a work on steam railways, then you might well find shops that specialise in transport issues will take the book. If you publish a book about a particular football club, then you might be able to get the club's shop to stock some copies.

Thirdly, there is advertising and PR. Advertising should be treated with caution – display advertising in magazines can be quite disappointing sometimes. Getting copies reviewed in specialist magazines can help, as can classified adverts in such magazines (many people find that although the sales are not great through the classified sections of specialist magazines they can be profitable).

And finally, there is a method which many publishers forget but which one of the publishers who uses Direct POD has used with some success and which we would recommend that you look at.

Our client published a novel which focussed on Arsenal football club – a book obviously intended for the supporters of the club. In order to sell the book the author set up a daily blog, recording the details of Arsenal, and with the usual discussions on the site day by day.

At the end of each article an advert for the book appeared, and the author also mentioned the book within many of the articles that appear on a daily basis.

After two years of blog writing the site gets over 170,000 individual readers a month – far more than most specialist magazines, and it is the main selling tool for the book. The author is busy writing a follow up novel, and uses the blog to try out ideas that will be incorporated in the book.

If you would like to know more about any of these aspects of selling, please do call me on 01604 882 588.

Environmental Policy

Our company recognises the importance of environmental considerations and will take these into account at all stages of business planning, development, production and marketing whilst continuing to produce high quality products to our customer's specification and satisfaction.

In particular the following objectives will be pursued as far as is reasonably practicable:

1. Conservation of energy. Through the appointment of our "energy champion" the company constantly seeks to reduce its energy wastage and overall usage .

2. Conservation of natural resources used in our manufacturing and despatch operations, particularly in the use of packaging. The company recycles over 90% of all waste paper and uses vegetable oil based inks and FSC accredited stock wherever possible. We will not accept products from Asian Pulp and Paper (APP) or Asia Pacific Resources International Ltd (APRIL) due to the unsound environmental practices of these Organisations.

3. Minimisation of waste in production and packaging.

4. Reduction of solvents in the workplace and atmosphere.

5. Provision of safe, clean and hygienic working conditions.

6. Reduction of noise levels to lowest levels practicable for employees and members of the public and the provision of ear defenders for general usage.

7. Safe storage and disposal of waste to meet "Duty of Care" requirements.

8. The responsible development, operation and maintenances of plant and associated processes to reflect the principles above.

9. In pursuit of our environmental aims both the Managing Director and Field Sales Executive drive hybrid cars which run predominantly on LPG.

Future-proof solutions

For planet-friendly business

Oce products that are clean and quiet

Since Océ products are an essential component in our working environment, it is important that they operate as cleanly and quietly as possible. At Océ, they set strict requirements regarding substance emissions and noise levels.

The discharge of vapours, gases and dust is strictly managed. When Océ equipment is installed in a print room with the volume and ventilation rate as defined in their requirements, concentrations of ozone, styrene and dust are very low. Océ use internal tooling and working methods to predict and reduce noise production. This ensures that our equipment operates as quietly as possible.

Future-proof solutions

For planet-friendly business

Products that last

Printing and copying equipment represent an inherent environmental burden. That is why the Océ lifecycle development programme considers environmental impact in the entire product cycle—from product development, manufacturing, use and reuse through to the processing of waste and residual materials. In fact, 'design for reuse' is one of their primary development criteria. Right from the start, they seek to make maximum use of materials and components. Océ ecological improvements include:

• Drastically extended lifetime of photoconductors to reduce waste

• Use of polypropylene toner bottles instead of high-grade plastic toner cartridges because disposal by clean burning is better for the environment than a cartridge take-back and reuse scheme

• The Océ 'upgrade before replacement' principle extends the lifetime of products, avoids waste and saves resources

• Océ has extended the lifetime of consumables

• Since many years, Océ has eliminated the use of specific hazardous materials such as benzene, cadmium, PBB and PBDE, PCBs and PCTs, and ozone depleting substances

• Océ continually works on eliminating the use of hexavalent chromium, mercury and lead in equipment

• Output trays for prints and copies are positioned at an ergonomically convenient height to reduce unnecessary bending

• Grips, handles and buttons are designed to be robust and recognisable for easy handling

• Built-in lifting aids in our wide-format and continuous feed printers make it easier to replace heavy rolls of paper

• Production printers have a sophisticated mechanism for exchanging highlight colour units

• The Océ CPS900 full-colour system won an award for its inventive and dust-free toner refill mechanism

Business operations that prevent waste

Océ seeks to reduce the quantity of waste produced by business and manufacturing facilities by optimising waste collection, separation, reuse and recycling, and by minimizing the amount of waste generated by production processes. All waste is disposed of in an environmentally responsible manner. We have taken steps to reduce emissions of organic solvent vapours—which contribute to the formation of photochemical smog and constitute a public health hazard— by optimising our manufacturing processes. For example:

• Océ does not produce any toxic waste

• Océ does not use CFCs, halons and substances that deplete the ozone layer in manufacturing

• Océ has reduced paper consumption during testing by 90% since 1995

• In 1990, Océ set up a dedicated asset recovery department that organises the flow of returned goods and maximizes their reuse value

Information that promotes awareness

Our approach is based on openness and objectivity. We keep our staff, customers and other stakeholders informed about environmental matters. Information is published on our corporate website www.oce.com and can be found in the annual Océ Sustainability Report. We advise our customers how to manage their document flows in an environmentally considerate way.

United Nations Global Compact

Océ pursues its sustainability and responsibility aims in an open and systematic manner. For that reason we have subscribed to the United Nations Global Compact, and we report along the guidelines of the Global Reporting Initiative.
We participate in international and national environmental regulatory commissions and advisory bodies to ensure that our manufacturing processes and products comply with all environmental requirements. Océ asks suppliers and business partners to adhere to these principles as well.

Printing for Professionals

Océ helps the people who make our world. Companies everywhere use Océ technical documentation systems in manufacturing, architecture, engineering and construction. High speed Océ printing systems produce millions of transaction documents each week, such as bank statements and utility bills. And in offices around the world, people use Océ professional document systems to keep the wheels of business and government turning. Océ is also at work in publishing on demand, newspaper production and wide format colour for spectacular display graphics.

DIRECT-POD

THE PRINT ON DEMAND COMPANY

Standard conditions of contract

('the Company' means Direct-POD Ltd)
In these conditions, 'electronic file' means any text, illustration or other matter supplied to the printer in digitised form on disk, through a modem, or by ISDN or any other communication link.

1. Price variation Estimates are based on the Company's current costs of production and, unless otherwise agreed, are subject to amendment on or at any time after acceptance to meet any rise or fall in such costs.

2. Tax The Company reserves the right to charge the amount of any value added tax payable whether or not included on the estimate or invoice.

3. Preliminary work All work carried out, whether experimentally or otherwise, at the customer's request shall be charged.

4. Copy A charge may be made to cover any additional work involved where copy supplied is not clear and legible.

5. Electronic files
(a) It is the customer's responsibility to maintain a copy of any original electronic file.
(b) The Company shall not be responsible for checking the accuracy of supplied input from an electronic file unless otherwise agreed.
(c) Without prejudice to clause 15, if an electronic file is not suitable for outputting on equipment normally adequate for such purposes without adjustment or other corrective action the Company may make a charge for any resulting additional cost incurred.

6. Proofs of all work may be submitted for customer's approval and the Company shall incur no liability for any errors not corrected by the customer in proofs so submitted. Customer's alterations and additional proofs necessitated thereby shall be charged extra. When style, type or layout is left to the Company's judgement, changes there from made by the customer shall be charged extra.

7. Colour proofs Due to differences in equipment, paper, inks and other conditions between colour proofing and production runs, a reasonable

variation in colour between colour proofs and the completed job will be deemed acceptable unless otherwise agreed.

8. Variations in quantity

(a) Every endeavour will be made to deliver the correct quantity ordered, but estimates are conditional upon margins of 5 per cent for work in one colour only and 10 per cent for other work being allowed for overs or shortage (4 per cent and 8 per cent respectively for quantities exceeding 50,000) the same to be charged or deducted.

(b) Continuous Stationery - Every endeavour will be made to deliver the correct quantity ordered but quotations are conditional upon the following margins being allowed for overs and shortages (measured in fold depths), the same to be charged or deducted:

For quantities below 10,000, or where special papers or special features are required: 10% margin Single part or one process work: 10,000 to 50,000 - 5% margin, over 50,000 -4% margin Multi-part, multi-unit or multi-process work: 10,000 to 50,000 - 10% margin, over 50,000- 8% margin

9. Delivery and payment

(a) Delivery of work shall be accepted when tendered and thereupon, or if earlier on notification that the work has been completed, payment shall become due thirty days after the end of the month in which the company raises the invoice unless otherwise agreed.

(b) Unless otherwise specified the price is for delivery of the work to the customer's address as set out in the estimate. A charge may be made to cover any extra costs involved for delivery to a different address.

(c) Should expedited delivery be agreed an extra may be charged to cover any overtime or any other additional costs involved.

(d) Should work be suspended at the request of or delayed through any default of the customer for a period of 30 calendar days the Company shall then be entitled to payment for work already carried out, materials specially ordered and other additional costs including storage.

10. Ownership and risk

(a) The risk in all goods delivered in connection with the work shall pass to the customer on delivery.

(b) Goods supplied by the printer remain the Company's property until the customer has paid for them and discharged all other debts owing to the Company.

(c) If the customer becomes insolvent (as set out in clause 16) and the goods have not been paid for in full the Company may take the goods back and, if necessary, enter the customer's premises to do so, or to inspect the goods.
(d) If the customer shall sell the goods before they have been paid for in full he shall hold the proceeds of sale on trust for the Company in a separate account until any sum owing to the printer has been discharged from such proceeds.

11. Claims Advice of damage, delay or loss of goods in transit or of nondelivery must be given in writing to the Company and the carrier within three clear days of delivery (or, in the case of non-delivery, within 28 days of notification of despatch of the goods) and any claim in respect thereof must be made in writing to the printer and the carrier within seven clear days of delivery (or, in the case of non-delivery, within 42 days of notification of despatch). All other claims must be made in writing to the Company within 28 days of delivery. The Company shall not be liable in respect of any claim unless the aforementioned requirements have been complied with except in any particular case where the customer proves that (i) it was not possible to comply with the requirements and (ii) advice (where required) was given and the claim made as soon as reasonably possible.

June 2009 | Version 1.1
Updates:
www.direct-pod.com/working.html
Page 2 of 2
June 2009 | Version 1.1
Updates:
www.direct-pod.com/working.html
Direct Print On Demand Limited
www.direct-pod.com • 01604 882588 • sales@direct-pod.com

12. Liability
(a) The Company shall not be liable for indirect loss or third party claims occasioned by delay in completing the work or for any loss to the customer arising from delay in transit, whether as a result of the Company's negligence or otherwise.
(b) Insofar as is permitted by law where work is defective for any reason, including negligence, the Company's liability (if any) shall be limited to

rectifying such defect. Where the printer performs its obligations to rectify defective work under this condition the customer shall not be entitled to any further claim in respect of the work done nor shall the customer be entitled to treat delivery thereof as a ground for repudiating the contract, failing to pay for the work or cancelling further deliveries.
(c) Nothing in these conditions shall exclude the Company's liability for death or personal injury as a result of its negligence.

13. Standing material
(a) Metal, film and other materials owned by the printer and used by him in the production of type, plates, film-setting, negatives, positives and the like shall remain his exclusive property. Such items when supplied by the customer shall remain the customer's property.
(b) Type may be distributed and lithographic or photogravure film and plates, tapes, disks or other work effaced immediately after the order is executed unless written arrangements are made to the contrary. In the latter event, rent may be charged.
(c) The Company shall not be required to download any digital data from his equipment or supply the same to the customer on disk, tape or by any communication link unless written arrangements are made to the contrary.

14. Customer's property
(a) Customer's property and all property supplied to the Company by or on behalf of the customer shall while it is in the possession of the printer or in transit to or from the customer be deemed to be at customer's risk unless otherwise agreed and the customer should insure accordingly.
(b) The Company shall be entitled to make a reasonable charge for the storage of any customer's property left with the Company before receipt of the order or after notification to the customer of completion of the work.

15. Materials supplied by the customer
(a) The Company may reject any film, disks, paper, plates or other materials supplied or specified by the customer which appear to him to be unsuitable. Additional cost incurred if materials are found to be unsuitable during production may be charged except that if the whole or any part of such additional cost could have been avoided but for unreasonable delay by the Company in ascertaining the unsuitability of the materials then that amount shall not be charged to the customer.
(b) Where materials are so supplied or specified, the Company will take every care to secure the best results, but responsibility will not be accepted

for imperfect work caused by defects in or unsuitability of materials so supplied or specified.
(c) Quantities of materials supplied shall be adequate to cover normal spoilage.

16. Insolvency Without prejudice to other remedies, if the customer becomes insolvent (namely, being a Company is deemed to be unable to pay its debts or has a winding up petition issued against it or has a receiver, administrator or administrative receiver appointed to it or being a person commits an act of bankruptcy or has a bankruptcy petition issued against him) the printer shall have the right not to proceed further with the contract or any other work for the customer and be entitled to charge for work already carried out (whether completed or not) and materials purchased for the customer, such charge to be an immediate debt due to him. Any unpaid invoices shall become immediately due for payment.

17. General Lien Without prejudice to other remedies,in respect of all unpaid debts due from the customer the Company shall have a general lien on all goods and property in his possession (whether worked on or not) and shall be entitled on the expiration of 14 days' notice to dispose of such goods or property as agent for the customer in such manner and at such price as he thinks fit and to apply the proceeds towards such debts, and shall when accounting to the customer for any balance remaining be discharged from all liability in respect of such goods or property.

18. Illegal matter
(a) The Company shall not be required to print any matter which in his opinion is or may be of an illegal or libelous nature or an infringement of the proprietary or other rights of any third party.
(b) The printer shall be indemnified by the customer in respect of any claims, costs and expenses arising out of any libellous matter or any infringement of copyright, patent, design or of any other proprietary or personal rights contained in any material printed for the customer. The indemnity shall include (without limitation) any amounts paid on a lawyer's advice in settlement of any claim that any matter is libellous or such an infringement.

19. Periodical publications A contract for the printing of a periodical publication may not be terminated by either party unless 13 weeks notice in writing is given in the case of periodicals produced monthly or more

frequently or 26 weeks notice in writing is given in the case of other periodicals. Notice may be given at any time but wherever possible should be given after completion of work on any one issue. Nevertheless the printer may terminate any such contract forthwith should any sum due thereunder remain unpaid.

20. Force majeure The Company shall be under no liability if he shall be unable to carry out any provision of the contract for any reason beyond his reasonable control including (without limiting the foregoing): Act of God; legislation; war; fire; flood; drought; inadequacy or unsuitability of any instructions, electronic file or other data or materials supplied by the customer; failure of power supply; lock-out, strike or other action taken by employees in contemplation or furtherance of a dispute; or owing to any inability to procure materials required for the performance of the contract. During the continuance of such a contingency the customer may by written notice to the Company elect to terminate the contract and pay for work done and materials used, but subject thereto shall otherwise accept delivery when available.

21. Law These conditions and all other express and implied terms of the contract shall be governed and construed in accordance with the laws of England.

Notes

Notes

Notes

Notes

Notes

Notes

Notes

Notes

Notes

Notes

Notes

Notes

Notes

Notes

Notes

Notes

Notes

Notes

Notes

CALL US FOR A **FREE** QUOTATION **01604 882588**

Notes